Knowledge is "Real Power"

Introduction to Power Quality

by Mark A. Shirah, CPQ

Knowledge is "Real Power"
Introduction to Power Quality
By Mark A. Shirah
1. Power Quality and Reliability
ISBN: 979-8-9852978-0-5 (paperback)
979-8-9852978-1-2 (e-Book)

Cover and interior design by Darlene Swanson, Van-garde Imagery, Inc.
Edited by Karin L. Nicely Lord

Printed in the United States of America

Seren Publishing Co., Inc.

SEREN
—PUBLISHING—

Ocala, Florida
www.serenpublishing.com

Preface

THIS POWER QUALITY REFERENCE BOOK is being written at the request of many linemen, engineers, engineering techs, troublemen, key account managers, safety officers, customer service reps, meter techs, and many others I have taught in my **Power Quality & Reliability** classes or who I have met throughout the country during my career.

In this introductory book, my goal is to explain very complex subjects in layman's terms, and I am thankful God has blessed me with the ability, opportunity, and platform to share this information. Additionally, my real-world case studies will illustrate various situations, problems, and resolutions.

My second book will dive much deeper into the technical side of power quality, where in most cases the problems and solutions are found, and will present more in-depth case studies. Also, once you have read this *Introduction to Power Quality*, the second book will be that much easier to understand.

I have been in many meetings and seminars with those having varied levels of knowledge, experience, and skill in this field, and I have looked around the room and could tell that many were lost in the complexity of the subject matter.

My basic approach to explaining power quality and reliability has been extremely popular over many years, and I hope you will gain some useful knowledge from my simplified approach to explaining this complex subject matter, as well.

Table of Contents

CHAPTER 1

Safety Basics

SAFETY IS THE MOST IMPORTANT aspect of most any job. Although your level of necessary caution will depend primarily upon your particular job duties, please take safety seriously all the time. Ultimately, your safety is in your hands, and there is no room for shortcuts in this area.

I know the fire-resistant (FR) shirts and pants are hot, especially in the southern states during the summer, and the electrical hazard (EH) rated boots and overshoes are heavy and hard to walk in at times, but they are necessary for a reason: TO KEEP YOU SAFE!

The goal is to return home every day in the same condition you left that morning. I know from experience what it is like to leave the parking lot at the end of a day while your friend and co-worker's truck is still there—because he will not be driving it home ever again.

I have lost two friends to linework over the years, and it is something that will stay with me forever. Linework is an extremely dangerous position which takes a very special person. I could go on and on about safety, but my advice is that you play a major role in your own safety; you must take ownership of it and realize whether you return home each day depends an enormous amount upon how seriously you take your own safety practices.

- Test your high-voltage gloves daily.

- Take time to discuss your job with your crew and what each person is going to be doing.

- Ask if each person understands their role; do not simply assume that they understand it.

- Talk about possible safety issues for that specific job, and ask for input from your crew.

- Let your crew know you are available to answer their questions at any point of the job.

- If in doubt about anything, STOP! Discuss the issue until everything is understood by everyone.

- If you are too busy to follow the very basic safety guidelines above, you are too busy to do the job safely and should not proceed.

CHAPTER 2

Power Quality Overview

POWER QUALITY IS AN EXTREMELY broad subject that has a multitude of different parts. Every aspect of a Utility, and every person working with the Utility, plays a part in providing proper power quality and reliability to their customer base.

Power quality has changed drastically through the years due to rapid changes in technology. Those rapid changes are exactly why this book is being written.

With changes in technology come certain advantages. Some of those advantages are variable frequency drives, better known as VFDs. Many customers use these devices for their efficiency and ability to control production-line speeds along with many other aspects of manufacturing.

Another greatly used item is a programmable logic controller, better known as a PLC. The logic controller, in many cases, is somewhat of the brain of the VFD, and other devices, as it is programmed to initiate different tasks for many different devices. These are just a couple of devices that are quite common with the newer technologies many businesses are using.

As with most things that bring advantages, newer technologies usually bring some disadvantages, as well. For VFDs and

PLCs, these disadvantages include nonlinear loads. By this, I mean they do not produce a normal sinewave signature.

The result from nonlinear loads is harmonic content, which in some cases can cause excess heating, reverse torque, data errors, and many other issues that must be monitored to verify they are not a problem. However, harmonic content should not be a concern when you understand what to expect and how to resolve any issues that may arise.

We will dig much deeper into the nonlinear subject matter in later books, but for now, just be aware that some issues can arise when using certain newer technologies.

My concerns are that many utilities are not keeping up with the increased learning curve the newer technologies present to their workforce.

Throughout this manual, we will discuss numerous topics, such as cause and effect along with possible resolution or mitigation techniques.

Learning mitigation techniques will take time, but the basics can be learned in a rather short timeframe. The more complex aspects of power quality and reliability, including how to read sinewave signatures, can take many years to perfect, but you must start somewhere.

My advice is to learn to look at power quality and reliability problems from a different perspective. Most people in the electrical field look at an issue from a voltage or current standpoint from the Utility or customer only.

Learn to look at a much broader picture; look at possible causes from both the Utility and the customer, as you normally would, but also look deeper. Look at possible environmental issues and other potential factors.

Many times, you may need to peel away prospective problems, as you would peel away the layers of an onion, to get to the real culprit. When you realize that you need to consider much more than just voltage and current, you will be amazed at what other possible causes you will become aware of.

Soon, looking at issues much more in depth and considering all aspects will become second nature.

In the following chapters, I will provide several of my real-world case studies with details of causes, effects, and how I resolved the issue.

When investigating a power quality concern, please do not get in a hurry, as I mentioned above. Take your time and look at all factors involved. All of this will make sense the more you read further into the book.

Do not make more of the issue than what it needs to be: KEEP IT SIMPLE!

If you later find it is necessary to get more involved, then do so, but not until all the basics are verified. Remember that over 65% of all power quality issues are grounding- or connection-related problems.

Now, let's get started with learning the basics of power quality and reliability, as *Knowledge is "Real Power."*

Chapter 3

Where Do I Start My Investigation?

When you are starting your power quality or reliability investigation, the most important aspect is knowing the right questions to ask. There are many, many factors that can change the way you will conduct the investigation.

First, you need to identify whether the issue is a Utility problem or customer problem.

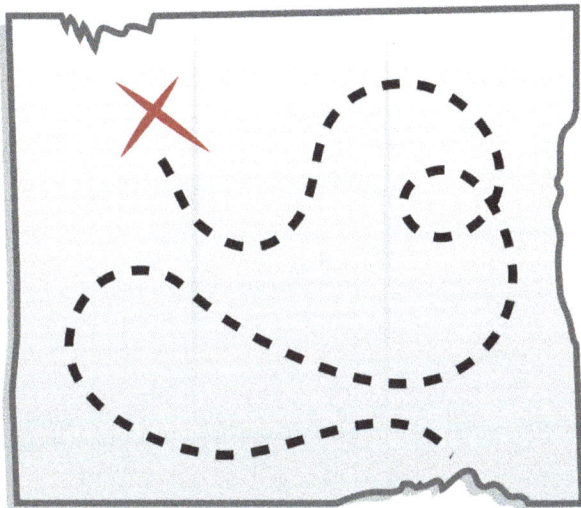

Next, if only one customer is stating issues, ask if the entire home or facility is affected. Ask them to explain what they are seeing or what equipment is being affected, as this will tell you how you need to proceed.

If the customer states that only one device or one area of the home is affected, then you know this should be an issue isolated to just their facility, and you need to have a troubleman verify good connections and proper voltage to the top side of the meter from the Utility.

If proper voltage is being provided from the Utility, then instruct the customer to have an electrician check for possible issues on their side.

If the Utility troubleman finds poor voltage or bad connections on the Utility side, repairs should be made and voltage should be corrected, thus resolving the issue.

If the issue is more widespread—for example, if the customer and neighbors are experiencing the same problems, such as lights going bright and dim—this is a good indicator that a bad neutral may exist, and this needs to be resolved ASAP.

However, if the customer and neighbors are experiencing a complete outage, this may be a blown transformer fuse or possibly a blown division fuse feeding that section of primary line. This will require a troubleman to investigate to find the outage area and determine the possible cause.

Once the line has been ridden out by the troubleman to verify the possible cause, and to make sure no lines are down and all factors involved are found to be safe to re-energize, a new fuse can be installed and the line can be re-energized.

The troubleman should record the cause if a cause was found, as this will help engineering to determine if any mitigation or other upgrades may be needed.

If a cause was not found, then the troubleman should provide as much information as possible, which may include time of day, a possible overload issue, weather conditions, and a general description of the area (tree canopy, vines, etc.).

This will provide data that can be very useful to help identify possible problems or justify upgrades if other outages occur in the same general area.

If the outage is much larger with many more customers involved, the issue may be a circuit outage, where the entire circuit is locked out, which usually affects a couple thousand customers.

Check Utility breaker data to determine whether the breaker has locked out at the substation. If so, have a troubleman or crew respond and ride the circuit to determine cause.

Many times, if a circuit is locked out and the problem has been found, crews are dispatched to perform load switching to restore power to as many customers as possible and to isolate the problem area.

If the issue is not a complete circuit lockout, and the customer states that the lights blinked and came back on, then chances are the recloser on the circuit recognized a fault and tripped and returned power, as the recloser is designed to do.

Often, this will cause customer equipment to trip offline, mainly in industrial or commercial facilities that have VFDs or

PLCs, robotics, or other types of equipment. Usually, they can be restarted and will return to service.

If you have access to Utility data, look at the reclosing and breaker events for the customer circuit. If an issue exists, follow up as needed to determine cause of event. This will include field investigations and digging through data to determine whether there may be an issue in the general area that may need repairs or upgrades to enhance reliability.

You will be amazed at how quickly you can determine if the issue is customer-side or Utility-side just by asking the correct questions.

The aforementioned scenarios are only a couple of possible events that may take place.

CHAPTER 4

Use Your Available Resources

SOME OF THE MOST VALUABLE assets at any Utility or facility are your co-workers. Your co-workers are a valuable resource that most people do not rely on and often even overlook as a possible resource.

Most people enjoy helping others, and although we are all busy, we each can take a moment to assist a co-worker, as we are all working together to produce a superior product to our customer base.

Remember that you are also a valuable resource to your co-workers. If they can pick up a phone and call or send a text to a co-worker for a quick answer to a question, it can save a lot of time and effort in research or investigations.

Many people do not feel comfortable asking co-workers for information because they are concerned they may be bothering them if they do so. I have asked this question in my **Power Quality & Reliability** classes, and guess what: the overwhelming majority of attendees would prefer their co-workers to use them as a possible resource, as they would then feel more comfortable asking them, in turn, when they need to ask a question. In my opinion, this can be quickly resolved by just asking your co-workers in your area and in other departments if it would be okay for you to call them from time to time as a direct resource. Then let them know they have a direct resource in you, as well.

I have seen this make friendships that otherwise may not have existed, so I see this as a win, win, win, win.

It's a win for each employee; it's a win for the customer, as power quality and reliability could be improved; and it's a win for the Utility, as employee engagement, interaction, and moral are improved.

I know it is difficult to take the first step toward developing the new resources, but believe me, it is worth it in the long run. And in more cases than not, they will be glad to have a new resource, too.

I look forward to hearing about how this strategy works for you, so please email me and update me with your progress. mark@cardinalpq.com

Chapter 5

Proper Outage Reporting Basics

Incorrect reporting is one of the biggest mistakes that most utilities make. In my opinion, there should *never* be an outage listed as "unknown"—EVER!

The average Utility has between 15% and 25% of their total outages listed as "unknown." There are some utilities that describe unknown outage as undetermined, but it all means the same thing: nothing was found as a cause of the outage.

Many utilities list outages as due to lightning or storms. These are just unknown outages in bad weather! Many times, outages are blamed on storms or lightning just because it's an easy way to assign a cause and move on. Instead, provide some information so there can be follow up to try and determine a cause and possible mitigation.

Over 20 years ago, I was in a round-table discussion about power quality and reliability with several other Utility linemen, troublemen, line crew foremen, and managers. I mentioned that I had come up with a way to improve reliability by resolving unknown outages. Of course, they all thought I was crazy and looked at me like I had lost my mind.

Before I could explain myself, they each took a turn providing me their thoughts.

The first person asked, "How do you find the cause of an unknown outage when you have no idea where to look or what to look for?"

The second person stated, "We have tried to identify unknown outages, and the result was a lot of man-hours wasted with no results."

The third person stated, "I wish we could afford to implement a plan to resolve our unknown outages, but when we spend money to resolve outages, we need to know we are spending money wisely and [actually] resolving outages, and not just in hopes of resolving outages."

After they had given me their input, we took a break, and I quickly wrote down their comments. I knew if I could convince this group that my strategy works, anyone at any Utility would listen to me.

When we all returned to the meeting, I explained my recent success with resolving unknown outages, and the group was glued to my every word.

I explained a general overview, as I will explain here, but with a few more moving parts. Overall, though, it is a simple plan.

I stated that proper reporting is the key to the entire plan. For example, when a line crew, troubleman, or whoever responds to an outage area finds no cause, they should *never* report the outage as "unknown." They should report a "possible cause" by providing every detail possible, including the date, weather conditions, and a general description of the area (such as trees, vegetation, possible loading issue, pole number, address, etc.).

That report is then provided to either system control or engineering, and preferably both, as this will ensure that someone has the record. Then Engineering or the power quality and reliability person will list the outage information on a map with a number. In a corresponding log, then, will be a number with details pertaining to the possible cause of the outage.

When you repeat this process for each possible-cause outage, at the end of the month, you can look at the map for possible patterns. I have seen, in most cases, that several of the "unknown" outages will be grouped in a general area, many times having the same possible cause: possible trees, possible squirrel, possible lightning, and so on.

This is when you go into the field to look for possible causes and identify possible needed upgrades to resolve the unknown outages.

If there are a lot of trees, or a thick tree canopy, maybe more animal guards or insulated jumpers are needed, or maybe Hendrix cable for falling limbs from overhang.

If the area is clear and no possible issues are seen, look for possible fuse coordination issues or overloading.

The most important aspect in this entire process is the complete cooperation of your line crew and system control staff. They must believe in this concept and take the time to provide the details needed that will provide you the necessary data.

In many cases, overtime will be reduced, and many will see this as a negative, so rewarding your employees when you see the reliability improve is key to the lasting success of this mitigation model.

I have used this approach for many years, as have many other utilities, which have stated that their reliability has increased. Several have reported that their overtime was reduced as well.

If your Utility does not have a power quality and reliability program, please contact Cardinal Power LLC with any questions.

CHAPTER 6

Everyone Has Special Skills

ONE OF THE BIG MISCONCEPTIONS is that one position
is more valuable than the other. Many may disagree with me,
and that is okay, but I have always felt everyone has their own
special skill set.

This is especially true at a Utility, because if you take away
any function, the entire house of cards weakens very quickly.

An example would be: if you remove your receptionist or
secretary and leave all the paperwork for linemen, managers, or
a wide range of others to do, what do you think would happen?
That is right—complete chaos in most cases.

Let's consider if all the engineers were to leave; who would
understand all the wind loading on the poles or perform all of
the math calculations or run the specialized CAD programs?

What about linemen or customer service reps? What if
someone else, who did not have the specialized skills to properly
perform those duties, had to try and do those jobs? Obviously,
it would not be a good situation.

I think you get my point: each of us has our own skill set,
and each of us plays an important role. Some may make less

money than others, but that does not mean their role is not important.

The reason I state this is due to an employee in the customer service department asking me once if I could call one of our linemen and get some further information about a customer for whom he had recently resolved a problem. The customer service rep needed the information from the lineman to provide proper records.

I told her I would be glad to call the lineman, but then I said, "I must ask, why do you want me to follow up with him?"

The customer service rep answered she was just a customer service rep, and she felt the lineman would not listen to her or might even get upset with her for bothering him.

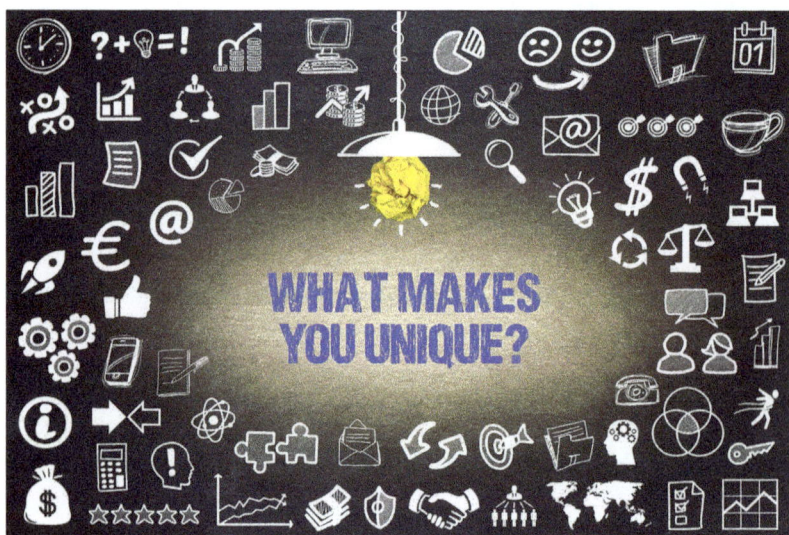

I invited her to talk with me in a small conference room. When we sat down. I explained that everyone plays an important role and assured her the lineman would have no problem talking with her and answering any questions regarding customers he had recently dealt with, because that is part of his job.

The customer service rep stated she did not understand all the things a lineman understands about power and the Utility.

I stopped her to explain that the lineman is not a customer service rep, either, and asked, "How do you think he would do if he had to try and do your job?"

The rep just smiled, as she understood what my point was. She called the lineman, and he did answer her questions. And the lineman even told her to feel free to contact him any time she had any questions he could assist her with.

She told me that had made her day. After that, she had a different outlook on the importance of her position and felt much more comfortable reaching out to co-workers.

Chapter 7

Listen with Respect, but Don't Be Led Down the Rabbit Hole

Listen to customers with respect, but do not be led down the rabbit hole. By this I mean: do not be led to believe everything a customer states is the way it really is, and no, the customer is not always right.

I know it is hard to believe, but customers may bend the truth from time to time to get a quicker response from the Utility or to get what they want.

Most customers have very limited knowledge of how a Utility system works, much less what repairs are needed when the Utility system does not work.

How many times have you had a customer call in to report a wire down and find it is actually a telephone or cable line? A leaning pole that is "about to fall," but is just raked back a little and not an issue. I would much rather a customer call in these types of complaints than to not call them in and have it turn out to really be a primary line on the ground or a pole in bad shape that *does* need attention.

My point is to not discredit customer concerns, even though they may not describe issues the way you or I, having many years of Utility experience, would.

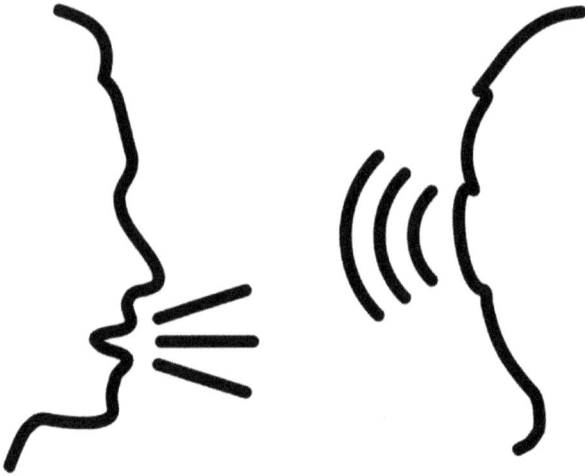

I recall an issue with a customer who frequently called in voltage problems and was well known in the customer service area for calling in at least once a month to have their voltage checked.

Each time, a troubleman responded and verified proper voltage. I had even installed a recorder at the meter and left it there for several days to get a good profile of Utility voltage and customer load. No issues were detected on the Utility side, and all looked normal on the customer side as well. The customer continued to call in about once a month for a very long time—the best I can recall is that it must have been a couple of years.

The same customer once called in a voltage complaint after a storm had passed through and many outages were in progress. The trouble truck quickly recognized the location and customer name and decided to put them at the bottom of the list, as he had been there many times before and had found all was good on the Utility side.

The customer called in again about an hour later and stated that something was wrong as she smelled smoke.

The troubleman received the message from the system control operator and quickly responded when he learned of the smoke odor. When the troubleman arrived, he was surprised to see fire trucks on the scene and smoke rolling from under the eaves of the home.

The troubleman immediately cut power to the home so the fire fighters could put water on the fire. The fire was quickly put out, and nobody was hurt. The troubleman later found where a limb had fallen and broken the neutral, thus allowing up to 240 volts on one leg of the 240-volt triplex service (two 120-volt hot legs and one neutral).

I have had customers call and tell me their transformer had blown up. I thought that, most likely, just a transformer fuse had blown, as when a fuse blows it sounds like a shotgun blast, but you always investigate that kind of report, regardless.

I have also had customers state that we "might" have an issue at a location, but once we arrive there, we find the pole is on fire, as lightning had hit the pole and knocked the top out of it, and the customer was getting partial power.

My point in this chapter is this: always listen with respect, as you really don't know what the issue is until you get there. And regardless, you must respond as a responsible Utility.

CHAPTER 8

Basic Calculations

BELOW ARE A FEW BASIC calculations that will help to identify possible loading issues. The below calculations are for a 12,470 system.

Please keep in mind that these are nameplate ratings of the transformers, as a transformer can be pushed well over its rated nameplate for several hours if it has a good amount of cooldown time.

To obtain full nameplate amps of a 120/240-volt single-phase transformer, multiply 4.17 by the listed kVA.

Example: 4.17 x 50 kVA = 208 amps

To obtain full nameplate amps of a 120/208 three-phase transformer, multiply 2.78 times the listed kVA.

Example: 2.78 x 3 - 75 kVA transformers = 624.5 amps

To obtain full nameplate amps of a 277/480 transformer, multiply 1.2 times the rated kVA.

Example: 1.2 x 3 - 250 kVA transformers = 902.1 amps

The above calculations are good for pad mounted transformers as well.

Calculating the loading on a delta bank is a little more involved. Remember that an open delta involves two transformers: a lighting transformer (usually the larger transformer) and a power transformer. The lighting transformer carries all of the single-phase load as well as 50% of the three-phase load.

The power transformer carries 50 % of the three-phase load only.

To figure full nameplate amps of an open delta: 120/240/-208 hi leg = 4.17 x the single-phase load and 2.41 x the three-phase load.

Open deltas should be used in specific places that have small amounts of three-phase load, with a good amount of single-phase load. Motors larger than 20 hp should require a closed delta.

Sewer lift stations are common for open deltas due to a known load that does not vary.

To figure full nameplate amps on a closed delta: 4.17 x the single-phase load, and 2.41 x the three-phase load.

The loading on a closed delta is: the lighting transformer (usually the larger transformer) carries ⅔ of the single-phase load, and the two wing transformers will carry ⅓ of the single-phase load, which equals ⅔.

For three-phase loading, all transformers carry ⅓ of the three-phase load.

Closed deltas are much better for locations that have bigger motors and large three-phase loads versus an open delta.

CHAPTER 9

Power Ride-Through Basics

BACKUP POWER HAS BECOME EXTREMELY popular over the years because of the changing technologies and the need for enhanced power quality and reliability. Most residential homes have what is known as an offline battery backup for their computers, which may be better known as the UPS (uninterruptable power supply).

These devices monitor the power feeding the device, which in most homes is a computer. If the voltage breaches the preset parameters of + or − 5% of nominal 120 volts (114–126 Volts), the battery backup is applied. You will normally hear a beep from the UPS when the backup is applied to alert you of the event.

The cost of the UPS is determined by the amount of time the battery is designed to power the device during the outage. An average cost of a one-hour offline UPS may be $59, whereas the same UPS device with a three-hour backup may cost $100.

There are two types of UPS: offline and online. The offline is what we just discussed and is in most offices and homes.

The online UPS is used where facilities have critical loads that cannot tolerate even the slightest power event. The online

UPS runs off battery power all of the time, and Utility power just keeps the battery charged.

During an outage, the device being fed by the UPS never knows that an event has happened; it is still running on the battery as usual, but it will only last as long as the battery is designed for.

The online UPS is usually much more expensive, and the cost also varies with the amount of backup time the battery can support.

UPS backup comes in many different sizes to accommodate various applications. For example, some entire facilities in Silicon Valley, California, use them to power critical operations that produce printed circuit boards. Even the smallest fluctuation in power can cause problems and cost hundreds of thousands of dollars.

Along with varying sizes, there are several types of backup power options. Most critical facilities, such as hospitals, police stations, nursing homes, and many others, are required to have backup power.

In most cases, this will be a preferred circuit, which they receive power from, as well as an alternate circuit just in case of an outage. There is usually a transfer switch that will transfer in about two seconds, allowing power to be restored.

In many cases, such as in hospitals, some equipment cannot ride through a two-second event, or even a one-second event, as one second is 60 cycles. Many robotic, VFD, or PLC devices can only ride through about four cycles (16.67 milliseconds per cycle).

This is where some locations use an online UPS device or a static transfer switch. A static switch transfers in less than four milliseconds and uses reverse voltage blocking technologies to make a seamless transfer without any fluctuation in voltage.

There is a wide range of options, but with the better options and seamless transfers, there is a higher cost involved.

There are pros and cons for each of the devices, of course. Battery backup has maintenance and battery replacement costs along with battery disposal fees. Diesel generators mean costs for maintenance, fuel, and fuel contamination maintenance. There is also the question of fuel availability during any type of disaster or event.

In addition, there are permitting issues regarding noise in many areas, where a generator is only allowed to run for a certain amount of time. There are also large footprints involved with the generator or the UPS.

The static switch requires very little maintenance and has very few moving parts. The cost of the static switch can be much more, but with all the maintenance issues, fuel cost, footprint, fuel availability, permitting, battery disposal, and other related costs, the static switch is worth considering as a possible option. Many people never consider all that is involved before deciding to use a generator or UPS.

One big misconception many customers and even some Utility personnel have is that if you have a backup circuit and a transfer switch, you will not be affected by outages. That is not the case. Most items will not ride through more than a few cycles, and motors can ride through a little longer due to the inertia from the rotating shaft, but only about six to eight cycles.

My advice is to look at all factors involved and choose your backup power wisely. I have seen many customers spend a lot of money, initially thinking they would have no outages, only to later find out certain parts of their plant would not ride through and a UPS was needed for that specific area.

I would recommend discussing your equipment with a power quality expert or someone specializing in power backup services, who may be able to offer you some advice and save a lot of time and money in the long run.

Chapter 10

Basic Harmonics

HARMONICS ARE A VERY COMPLEX subject that I will not dive into very much in this introductory book.

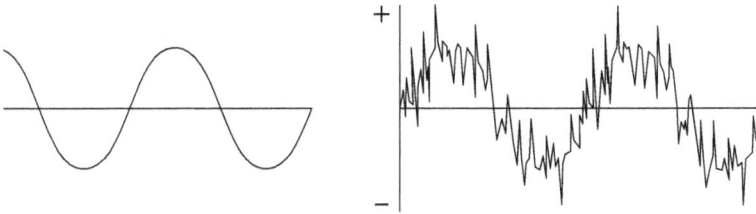

The main thing I would like you to remember about harmonics is: "They are only a problem if they are creating a problem."

Please read that sentence again. This is something I will mention in all my classes and in every book. Although harmonic issues can cause a lot of damage and can be a huge issue, harmonics exist in every service and voltage report.

The trick is understanding what a harmonic is. A harmonic is a multiple of the fundamental frequency. To put it in layman's terms, the first harmonic is 60 hertz/60 cycles, the second harmonic is 120 hertz/120 cycles, and so on.

The third harmonic is one that can cause a lot of issues, as it is considered a zero sequence harmonic. Therefore, instead of circulating in each phase conductor, the third harmonics from each phase conductor add in the neutral.

There are odd, even, and zero sequence harmonics. The odd or negative harmonics (second, fifth, eighth) can cause vibrations, reverse torque on motors, torque pulsations, and in severe cases, even broken shafts.

The even harmonics, also known as positive sequence harmonics (first, fourth, seventh), help motors turn in the proper direction and are not an issue in most cases.

Zero sequence harmonics, also known as triplen harmonics (third, sixth, ninth), add each phase conductor harmonic content in the neutral. Positive and negative harmonics circulate in each phase conductor.

Name	Fund.	2nd	3rd	4th	5th	6th	7th	8th	9th
Frequency, Hz	60	120	180	240	300	360	420	480	540
Sequence	+	-	0	+	-	0	+	-	0

This can cause overheated neutrals and sometimes even derate the capability of the transformer due to heating issues. The majority of harmonic issues are generated by non-linear loads such as VFDs or PLCs, robotics, etc.

Non-linear loads are caused when rectifiers are used, which changes the wave form and usual linear sinewave signature. A linear sinewave is a smooth normal voltage waveform. A non-linear voltage waveform can be very ugly and not even resemble a voltage waveform in many cases.

There are many different types of mitigation methods to reduce, eliminate, or cancel harmonics, but much needs to be considered before going to that point.

Harmonics can greatly affect your power factor, but if your Utility does not charge for a bad power factor, and you are not having any issues, why spend the money to mitigate something that is not a problem, as mitigation can be very expensive?

There are several vendors that make a very good living seeking out companies they know will have a poor power factor. They know the power factor will be poor by the type of equipment the companies are running.

Many times, vendors will provide a free voltage profile report so they can show you a very poor-looking voltage waveform. Then they say they can install power factor correction, provide you with a clean, very normal-looking waveform, and reduce your Utility bill. Many will claim up to 30% savings.

However, if your Utility does not charge a power factor penalty, you will not save any amount of money. This means that the Utility does not charge for reactive power.

Often, these vendors will even provide a small test and show where the voltage will increase, the amps will almost be cut in half, and the power factor improves, but they never show you the real power (kW). What you're billed for does not change. Most people think that if you reduce the amps, you will save money; that is not the case.

Years ago, a very large business was considering adding power factor correction capacitors to their facility. A vendor had convinced them the Utility was ripping them off and they could save over 30% of their bill each month if they were to buy the vendor's power factor capacitors.

The Utility key account manager and I met with the customer and the vendor, and we tried our best to inform the customer that no savings would occur if they installed the product.

The customer was not having any harmonic-related issues but had seen the poor-looking voltage waveform, and rather than seeking proper guidance, he took the vendor's advice.

The customer's normal monthly bill was usually about $300,000, so to say the customer would save at least $90,000 each month was a big statement.

The vendor proceeded to install the power factor correction at a cost of $230,000 for the customer. After a couple of months, we got a call complaining of no savings. We held a meeting and explained what we had recently mentioned.

The customer tried to contact the vendor, who could not be reached—and as far as I know, was never able to be reached.

Harmonics are a very complex issue that exists in any facility. In most cases they are not a problem, but if they are, you will know it.

I will get much more in depth with harmonics and their causes and effects in my next book, which will be titled *Knowledge is "Real Power" Level 2.*

Real-World Case Studies

THE FOLLOWING ARE A FEW of my real-world case studies. Some of these made me really think outside the box, look at the big picture, and consider all possible causes.

The Fish Farm

A FISH FARM, WHICH RAISED tropical fish, was complaining about burning up their new air conditioner circuit boards and destroying multiple single-phase motors that ran the inside irrigation pumps for the fish tanks.

The customer was very upset as this was a brand-new facility. The a/c was new, as were the motors that were failing. I quickly responded, looked at everything involved, and saw nothing out of the ordinary except eight of the nine very hot, single-phase, 1 hp motors for circulating water for the fish tanks. The first motor was maintaining a normal temperature.

I checked the voltage on sight, and all looked good. The customer received power from a closed delta transformer bank (120-120-208 high leg).

I installed a power quality recorder at the transformer and let it run for a few days. I then recovered the recorder and found good voltage and proper loading, but the harmonic content was extremely high, and there was a fault that looked very odd. I could tell from the load that it was the a/c unit.

I returned to the location and found that the a/c circuit board had failed again while the recorder was hooked up, but no other issues.

I had the customer contact his a/c person so I could discuss possible problems and maybe brainstorm with him. When the a/c person showed up, he was very defensive, saying there was no problem with the installation, etc., and I quickly calmed him down by stating that I just needed his expertise for possible ideas.

He was glad to help. He removed the cover of the a/c unit and the small cover of the failed printed circuit board area. That's when we discovered a lot of condensation on the board due to the warm, humid environment created by all the fish tanks.

This explained the failures of the a/c unit, as the customer was turning the a/c unit way up at night and would turn it down during the day for more comfortable working conditions. Thus, at night the condensation would build up, track across the board, and short it out as the droplets dripped across the energized board.

I was glad to have identified the a/c issue, but this did not explain the heating and failures of the single-phase motors of the same area.

I measured voltage, and all was very normal. I looked at the waveform with a Fluke 41B Power Harmonics Analyzer, and the waveform was very distorted, with very high harmonic content of many of the lower order.

This one had me scratching my head. No other items were having any problems: no lighting problems, no computer issues, nothing—only the motors.

I leaned back against one of the tanks and started thinking outside the box, asking myself, *What are all the things involved? Power—it's good. Ventilation—it's good. What could cause the heating for each motor other than the first motor? The only other factor is water, but how could that be connected?*

I looked at how they were connected and saw that nine motors were connected in line and set up as stagger start, meaning when the first one starts, then the second starts, and so on.

While looking at the plumbing of the water for the pumps, I realized there was no check valve installed between the motors to keep the water from being pushed backward when the first motor started.

Bingo! I knew I was on to something. If a single-phase motor is turning backward, and power is applied, it will continue to run backward.

What was happening when the first motor would start, without having a check valve in the water piping, was the water feeding through the second motor would be pushed backward. Then when it started, it would be started in the reverse direction.

All the motors were overheating due to this issue except the first motor, as it was turning in the proper direction. I had the customer contact his plumber, who stated the missing check valves were an oversight. He quickly installed them, and the issue was resolved.

As you see in this case, the Utility did not have an issue but was being blamed, and I fully understood the customer's concerns, as everything was new. The customer stated he was very grateful for the Utility involvement and resolution.

In many situations you may face, customers may not always give the Utility the benefit of being innocent until proven guilty.

In these cases, keep your cool, remain professional, and look at everything involved. Think outside the box and use your skills, tools, and resources to provide answers. These types of issues will provide you with an invaluable skill that will be a huge help with future complaints.

Voltage Flicker

VOLTAGE FLICKER IS USUALLY DESCRIBED as a very brief inconstant or wavering of light and should not be confused with voltage drop. Voltage drop is much different and will be discussed in my Level 2 book.

Some people are more sensitive to flicker than others, so voltage flicker is very subjective and heavily dependent upon human perception.

Voltage flicker is, in most cases, due to load starts such as a/c units or motor-starting inrush events. Voltage flicker is very normal, if you suspect it is outside of the normal Utility guidelines of + or - 5% of nominal voltage, issues need to be investigated.

The following issue was seen by the troubleman as a severe case of voltage flicker that needed further investigation.

I was asked to get involved when the issue was still present after all connections at the weatherhead, the open secondary, and the transformer had been changed. I was asked to meet with the customer to discuss everything involved and to be there late in the evening due to the voltage flicker issue only happening at specific times of the day.

When I arrived at the customer's location late one evening, I asked to see what the customer had been experiencing. The customer invited me inside the home, and I saw the lights flicker greatly. But rather than occurring randomly, which would be a neutral issue in most cases, this was showing a pattern of consistency. The customer was having no other issues with any other equipment.

Because the customer had a chandelier with clear bulbs, the flickering could be seen to a much greater extent, which exaggerated the issue. I asked the customer when the issue was happening. He replied that it only happened early in the morning or late in the evening. I asked if he had talked with any neighbors. The customer stated he had talked with the neighbor behind him but had found he had not been having any problems.

I explained to the customer that I would look at all factors involved and get back with him. I talked with the customer's neighbor and learned that he worked a lot and nobody was home most of the time.

The customer fed from a 37.5 kVA transformer which served three homes. There was about 75 feet of 2/0 open secondary to a primary and secondary dead end. The customer service was 1/0 triplex and was about 50 feet to the weatherhead. The neighbor's service was about the same distance and was the same size.

The customer's home was approximately 1,800 square feet, as was the neighbor's. The customer's and his neighbor's were the only two active homes; the other home was abandoned and the service had been disconnected. Looking at all factors involved, I found that the Utility voltage flicker should have been inside of guidelines.

The customer had been having this issue, which had started gradually and had worsened over time, for about three months.

I installed a recorder at the customer home and found very good stable voltage, but there was an odd amount of voltage fluctuation for about ten minutes each morning and again each evening. I found it even more odd that the voltage fluctuations occurred at almost the same time each day.

The recorder showed me this was not an issue from the customer's home, as there were no load changes to cause the events.

I then moved the recorder to the open secondary and saw the same events each morning and evening with the same timeframes. The customer's service connections and his neighbor's were side by side on the open secondary.

I was looking for possible causes and could not seem to put this puzzle together. I decided to show up on site the next morning at the time the events were happening each day to possibly discover something. To be on site was not within my normally scheduled timeframe for work, but it was necessary to see if anything was different. I informed the customer I would be there early the next morning.

I arrived on site at about 5:30 A.M. and just listened and watched while I leaned against the customer's fence, which backed up to the property of the neighbor he shared the open secondary with.

I had only been there a few minutes when I heard the neighbor's water-well pump turn on and shut off very quickly about every three to five seconds for approximately twelve minutes. When this started, the customer stepped outside and informed me that the flicker issue was happening.

This was all beginning to make sense. As the neighbor was about to leave for work, I caught up with him and asked if he knew his water-well pump was waterlogged. He explained he was aware of it but had not had a chance to fix it, and he knew it had been getting worse over the past couple of months.

I asked if he also normally turned on the water about 6:00 P.M., in addition to his early morning usage. He replied that after returning home at about 5:30 P.M., he would usually fix supper, wash dishes, and take a bath.

Aside from those times, nobody was home and no water would be used. I explained the issue was causing his neighbor to see the starting-and-stopping (inrush) events of the pump motor.

The neighbor stated he would get the issues resolved as quickly as he could. A few days later, the neighbor contacted me and stated all had been repaired. The damaged pump motor had been drawing three times the amps of a normal 1 hp pump motor, so it had been replaced.

This all made sense, and the customer stated the issue was resolved. It took going the extra mile to go in early and stay late to be on site when the events were happening. Without that, I am sure it would have taken longer to identify the issue.

Stray Voltage

INVESTIGATING STRAY VOLTAGE ISSUES CAN be like trying to find a needle in a stack of needles. Most stray voltage issues are very time consuming, and most are going to take more than a few moments to identify and resolve.

This comes up in my **Power Quality & Reliability** courses more than any other topic. I am sure that is why the Stray Voltage class is always stated to be the favorite of the many different classes.

The most-asked question by my attendees is: "Where do I start my search?"

This is a very valid question, but the answer depends on a number of factors. As with any voltage-related issue, there need to be certain questions asked: When did the issue start? What is being affected? Is the issue isolated to one home or location? What has changed, according to the owner's viewpoint, to possibly cause the issue?

With some of these questions answered, you should know when the issue began and whether it is isolated to one location. In most cases, I like to try and isolate the service to see if the issue is coming from the Utility or if the issue may be coming from a joint user, neighbor, or other possible source.

The biggest issue when dealing with stray voltage is *safety*. Stray voltage can be very dangerous, so use your gloves and be aware of all potential problems.

In Florida, a good amount of stray voltage issues will be found at swimming pools. Although there is an easy test to verify if the pool is properly bonded, this seems to be a well-kept secret as not many people seem to know how to perform the test.

First, you will need a voltage meter that will read ohm values. Next, you will need extra wire of a small gauge because you will be measuring from the service entrance ground to the case ground of your pool pump or any metal object connected to the pool, such as a ladder, handrail, etc.

If extra wire is needed for the test, just ohm out the wire and subtract that from the value of the measurement from the service entrance ground to the pool motor case ground.

Bonding Ohm Values

Ohm Values (ohms)	Description
0.0 to 2	Low ohm value indicates that the pool equipment is bonded.
100 to 10,000	Medium ohm value indicates that equipment has a loose connection.
Open or Infinite	Very high or open ohm value indicates that equipment is not bonded.

Notes:

1. The ohmmeter leads and any extra wire used should be zeroed out to obtain an accurate ohm measurement.
2. The medium ohm value of 100 to 10,000 ohms may indicate that pool ladder or handrail supports are not tightened properly. Tell the customer to clean and tighten the pool ladder or handrail supports.

The above chart provides ohm value data that will indicate if the pool has a proper bond, possibly has loose connections, or is completely without a ground bond. Below is another way to test. If one part of the pool has better grounding than other parts, a few tests around the pool area may help identify possible issues, such as coping stone problems, poor bonds of railing, etc.

I have always preferred to start at the problem location and work my way toward the Utility side by isolating the service. You always want to verify proper bonding of the service entrance ground with values of 25 ohms or less. If you see less than 1 ohm, you more than likely have a ground loop because a ground of less than 1 ohm is almost impossible. Most ground resistance testers will show less than a 1 ohm value as a code for a ground loop. Look in your user's manual to verify all codes when less than 1 ohm is being indicated.

All joint users, such as cable, phone, etc., MUST bond at the service entrance to avoid creating a difference of potential.

I have found this to be an issue many, many times throughout my career. This can cause severe issues for computers that have internet connections through their cable provider if the cable provider did not bond at the service entrance as the NEC (National Electric Code) requires.

Many times, the issue can be on the Utility side and caused by a poor neutral connection, as this is the Utility power return path back to the substation.

If the current sees a better path than the Utility, it tries to follow the path of least resistance. If the Utility has a poor connection in the neutral path, that path may not be the best path any longer, and stray voltage is created. In a three-phase system, the amount of unbalanced load will be seen on the neutral; therefore, verify the Utility primary is well balanced.

Wells, pools, and many other paths are very possible. Until the source can be tracked and resolved, Burndy has created a pool bonding kit that has been a huge success in resolving pool bonding problems. The kit has a nine-square-inch metal plate that attaches to the pool skimmer and has a place to connect a ground. This allows for the pool to be at the same potential as the pool pump case ground.

To explain in more layman's terms, what you are doing with pool bonding is keeping the water at the same potential as outside sources, so you are actually bonding the water.

From a Utility standpoint, there is a device that can block stray voltage coming from the Utility until the source can be tracked. It is called a Ronk Stray Voltage Blocker.

Many dairy farms are using these due to grounding issues they face.

A Ronk device is designed to be installed on the Utility side, normally just below the transformer. The device is designed to separate the primary and secondary neutrals during normal operation. However, during a fault, the primary and secondary neutrals will connect in less than two milliseconds, divert the fault to ground, and then separate again.

There are several different types of Ronk devices with a few different levels of voltage protection.

The previous pictures are of a Ronk device that was used on a stray voltage issue in the Orlando, Florida, area. This issue was later found to be caused by improper cable-provider grounding. The cable company was informed, and the grounding was corrected. The Ronk device was removed and thus ready for use in another area if needed.

This case study provides a very brief overview of how I approach and identify possible stray voltage issues.

Energized Neighborhood Fence

I WAS CONTACTED BY THE system control operator over the radio and asked to investigate an energized fence in a neighborhood. The controller stated this was a high-priority issue and that a trouble truck was being sent to the location to assist me as needed.

I was already in the field, working on a different issue, but stopped what I was doing and drove toward the priority issue. I was only a short distance away, so I was there in just a few minutes.

When I arrived at the location, several neighbors were inside their yards, stating they could not get out through their gate because they were getting shocked every time they tried to open it.

I quickly put on my high-voltage gloves, overshoes, etc., and checked voltage across the gate latch. When I separated the metal gate latch, I drew an arc. I measured 93 volts across the latch. About this time, the trouble truck arrived. I quickly had the trouble truck cut power to the street by opening the cutout for the division fuse feeding the street.

Once I had verified the power was off, I talked with some of the customers. Several were late for work and could not talk, but a few were able to discuss the problem.

The customers stated the issue had recently started and no other problems were occurring. I was amazed to learn how far the customers said the voltage issue was continuing down the street.

The entire block was being affected because the chain-link fences were all connected.

I looked at the entire area for possible issues and found nothing that stood out as a possible source. I informed the customers I was going to turn the power back on for a little while to try and track the source of the stray voltage.

Several more Utility people had arrived to assist me in my investigation as well as to watch and warn customers of the safety issue of the stray voltage on their fences.

The farther I went down the street, the less voltage I measured, so I returned to my initial 93-volt location and went the other way. My measurements went up as I went, and before long, I knew I had to be close to the source.

I stopped and just looked around for a few moments, searching for any possible cause. I saw nothing and continued in the same direction, measuring and looking. I then saw a small metal shed with a Romex wire coming from a hole in the side of it and draped over the fence.

I went to the location and *bingo*! I could see where the wire had a small burn mark on the top rail of the metal chain-link fence. I measured 118 volts at the location from the top rail of the fence. I could see that the jagged hole in the side of the building had nicked the wire insulation, which was contacting the fence.

Once again, I had the trouble truck lineman cut power to the area. I then went to the home and met with the owner, who stated he had just installed an above-ground pool and had run temporary power to the pool pump.

I made him aware that he was causing a severe safety issue for the entire neighborhood and the power would need to be disconnected to resolve the issue immediately.

The customer agreed and quickly disconnected the wire from his small disconnect in the shed and stated he would not reconnect it until it could be done safely and correctly as I had mentioned.

Earlier, I had called the Utility safety officer to assist me in finding the issue, so both of us were present when the customer stated he would not reconnect the power to the pool. The safety officer stated he would leave power to the home under one condition: that the power would not be reconnected feeding the pool pump under any conditions until it was done in the proper, safe manner.

The safety officer also stated that if he did reconnect it and the stray voltage returned, the power to the customer's home would be cut off and would not be turned on again until the county officials would inspect it to verify all was safe.

The customer agreed to the terms. As a Utility, we did not want to turn off the customer's power, especially with small children living in the home, as this would mean lost food in the refrigerator and freezer as well the family being without air conditioning and other aspects of comfort and need.

I had the lineman energize the neighborhood and then verified that no voltage was on the fence after the wire was removed. All was back to normal, at that point, with no stray voltage on the fences in the area.

The entire process of finding and eliminating the issue took about one hour once we were on site.

I had mentioned to the safety officer that I strongly suspected the customer, at some point, would reconnect the wire feeding the pool pump motor. He stated he really hoped I was wrong, as he did not want to cut power to the home with kids living there.

I also did not want to disconnect power to the home, but after talking with the customer, I just got the feeling he was not taking us seriously. And although he had never said it out loud, I believed his attitude was that nobody was going to tell him what he could and could not do on his own property.

Although we tried to explain the seriousness of the issue he was causing, I really do not think he ever grasped the full extent of it or the possible severe or life-threatening injury that could result.

A person's body language can tell you a good amount in some situations.

Unfortunately, I was correct in my thoughts. All of us from the Utility had left the area. I had not been gone for more than thirty minutes when I got a call that the voltage had returned on the fence again in the same area.

I called the safety officer to meet me at the problem location. When we arrived, the customer was inside his home, and when he saw us, he ran to disconnect the wire feeding the pool pump, but it was too late—we had seen him disconnecting what we had just instructed him not to connect until it was done correctly and safely.

The customer stated he just needed to get the pool clean and then he was going to cut it off until he could fix it correctly. We explained we would not take the risk of him reenergizing the neighborhood again and that his power was going to be cut off and would not be turned back on until all aspects of his electric were inspected by the county officials.

The customer was extremely upset and stated he was going to sue the Utility and have us fired. We reiterated the seriousness of the issue and reminded him we had given him one chance but he had failed to listen.

This was a very odd issue that could have had a bad ending if someone had been electrocuted or a pet had been harmed. I will never forget this case study due to the boldness of the customer to return power to his pool pump after being warned of the safety issues involved as well as the personal consequences he would face if he did.

Later that week, I was contacted by the city manager, who asked me to meet him in his office. When I arrived, the Utility general manager, assistant general manager, safety director, safety officer, and my direct manager were there.

I was asked to have a seat. I was told that the shed owner had filed a damage claim against the city in the amount of $10,000, and I was named in the claim as being responsible for the damages to the customer's home, spoilage of food, and lost time from work.

I then explained the entire process in detail. Suspecting this would happen, I had written down every piece of information from start to finish.

When I had finished my explanation—including how the events had occurred as well as the corrective actions I had taken concerning trying to work with the customer and not turn off the power to his home until I had had to for safety reasons—the city manager stepped over to where I was sitting. He asked our group to stand, and he went down the line and shook each of our hands and congratulated us on a job well done.

The city manager stated that not only had we acted with due diligence, we had also acted with compassion, and when compassion did not remedy the problem, we had acted with resolve. We each appreciated the recognition we received and were glad we had the support of upper levels in these circumstances.

It is my opinion that your workforce is only as strong as the upper-level support will allow them to be. It is quickly learned by employees what support they have and to what extent they can perform their job duties.

When you have support from upper levels, it is much easier to make decisions and feel comfortable in doing your job. There are also those times when you may not make the exact decision needed and discussions are needed to possibly learn how things may have been better handled if done differently.

We are all in a learning process, and by the time we have a lot of knowledge, it seems it's time to consider retirement. That is why I have decided to pass on what I've learned through many thousands of power quality investigations. I am always eager to get your thoughts or your input on any of the information in this book. Please feel free to email me at mark@ cardinalpq.com.

Radio Frequency Interference (RFI)

I RECEIVED A CALL FROM a local ham radio operator who stated he was unable to operate his radio on 40 meters (7.0–7.300 MHz). I could tell the customer was upset and very frustrated with not being able to operate his equipment as needed.

I scheduled a time to meet with him so we could discuss his concerns and I would have a chance to see the issue he described as being S-9 on his radio meter.

Because the customer was distraught and, I suspected, elderly, I scheduled the meeting as soon as I could. When I met with the customer, he quickly informed me he was a decorated WW2 veteran. I thanked him for his service, and we discussed his time in the war for a little while.

He explained he had been a radio operator in the Navy, and in those days, Morse code was used. He stated he was one of the few people left who could have a conversation in Morse code.

I found him to be very interesting, and I could tell he was checking out how much I knew about RFI, etc. I also could recognize that he was asking questions to see what kind of electrical background I had.

I told him my radio call sign of KJ4-IRD, and it seemed to calm him a little bit to know I also was a licensed ham radio operator—that is, until I told him I don't operate and that I had just gotten my license to understand more about his side of the radio.

He quickly let me know he had a lot of knowledge and, if I cared to listen, he would share with me some of his wisdom so we could resolve the issue quickly.

I listened with respect and looked around his "ham shack," the room housing his equipment. I was looking at several issues concerning improper grounding along with the broken knobs on his radio and a very old, humming fluorescent T-12 lighting ballast in close proximity to his equipment.

The customer stated that the issue was a Utility problem and said he knew it was on our big main line. I asked if he felt he had any issues on his side, and he quickly assured me there were no problems on his side and it would be a waste of our time to even look.

I asked if I could get a copy of the RFI signature waveform from his radio so I could use it to try and track the signature. I got the signature and thanked him for his time, telling him I would get back with him after I had looked at all involved factors and checked the general area.

The next morning, I took the oscilloscope, Yagi antenna, parabolic dish, and hotstick line sniffer in the field to investigate possible RFI issues in the customer's general area.

I soon found a couple of broadband issues on the very old 69 kV line in close proximity to the customer home (approximately 1/4 mile). I verified the sources as loose ground wire and staple, as well as ground wire touching a cross arm bracket.

I was able to find a tracking insulator and a noisy arrester on the distribution line. I had these issues resolved, and the RFI was reduced, but not to the customer's satisfaction. The insulator was not causing much of an issue, but it was resolved and may have helped a little. The arrester made a good difference when it was replaced.

However, the customer still was not satisfied. I told him I would check further and go from there. The customer stated he was going to call the FCC to get the issue resolved.

I told him that was his choice and I would be glad to work with anyone. If the problem was found to be on the Utility side, I assured him, it would be resolved in a timely manner.

I think this caught him off guard. It seemed as though he expected that when he mentioned the FCC, I would jump through hoops for him not to call them. But I had nothing to hide; I had acted with due diligence and had resolved several issues.

Over the next two weeks, I could find no other issues on the Utility side that would cause the customer problems. I talked with other ham radio operators in the general area and found they were not having any issues on 40 meters and was told they'd had some noise a few weeks ago, but it was now gone.

They each let me see for myself that 40 meters had S-2 to S-3, which is very normal and good. I had each of them swing their tower beams around and point them in the direction of the older gentleman's location, confirming the noise level had improved for each of them.

I thanked them for their time, but one customer followed me to my vehicle as I went to leave the last location. This customer told me he knew the older gentleman I had been working with and knew there were several issues on the customer's side. He explained the same grounding issues I had seen earlier, and I told him I had seen several issues myself and that I was planning to discuss those with the elderly radio operator the next morning. I had just wanted to make sure the Utility side was good before I mentioned anything on his side.

The next morning, I returned to the WW2 veteran's home and discussed possible issues on his side. He became upset at this and said I was just not wanting to fix anything on the Utility side, that I wanted to blame any problems on him.

I quickly explained that, on the contrary, I would go to great lengths to make sure he could operate his radio, if that's what he wanted to do in his retirement, because I knew he had earned it many times over.

The customer seemed to calm down a little and allowed me to check a few things. I quickly found he had 2.5 amps on his service entrance ground, with high noise content, which also was showing a code for a ground loop. His tower was ungrounded, but his radio was grounded to the garage-door metal rail.

I asked if he would turn off the humming fluorescent light above his radio to see if that made a difference. He refused, saying it would make no difference. I mentioned that I had done all I could on the Utility side and reiterated he did have several issues that needed to be investigated on his side.

I told the customer I would be more than willing to answer any questions an electrician or anyone else might have in order to help solve his issue.

I explained that several other ham operators in the same general area were having no issues on the same frequency, and I really felt that if he solved his internal issues, he would be in good shape concerning the RFI he was seeing.

The customer stated he was contacting the FCC and they would be in touch. I gave him my card and said I would be glad to work with them, that maybe I had overlooked something.

About a week later, I got a call from the FCC, asking me to meet with them at the customer location later in the week. I agreed to do so.

When I met the FCC agents at the customer's home, we discussed the issues I had found and resolved, and how the noise had gotten much better but had not been resolved to the customer's satisfaction.

I explained that the other ham operators were having no issues in the same general area at the same frequency.

After I had mentioned several of the problems the customer had on his side, one of the FCC agents stated that those issues needed to be investigated as they could easily be causing the problems.

We met back with the customer, who had gone in and had lunch while we discussed everything involved. When the customer returned, he asked if we had it figured out and when it was going to be resolved. One of the FCC agents stated he would like to do some checking on the customer's side.

The customer had a handheld AM/FM radio on the shelf, which the agent picked up and turned on. The noise was static, nothing but static. The agent then held it close to the fluorescent light, and it got worse. When he moved it farther away, the noise was much less.

The agent turned on the customer's ham radio and the noise level was about S-4 to S-5 and showing a good amount of static from the lighting ballast, but the noise level was much better than S-9, as it had been at the beginning. The agent then turned off the fluorescent light, and the noise was reduced to about S-3 to S-4.

I showed several grounding problems to the agents, and they agreed those issues could easily be the remaining sources of the customer's noise problem.

The FCC agents explained to the customer that he would need to have his issues resolved and that the Utility had acted with due diligence and had gone well beyond what many utilities would have done.

When the FCC agents left, I sat down with the customer and explained that I knew a few people who may be able to assist him with finding and resolving the problem he had with grounding, but I could not get involved from a Utility standpoint due to possible liability concerns.

The customer seemed defeated and sad, and that was the last thing I wanted, although he had needed to realize some issues did exist on his side.

He agreed to hire a local electrician I knew was very good at grounding issues. The electrician, who agreed to lower his normal rate as a good gesture for the decorated veteran, quickly found several of the issues I had mentioned as well as a couple I had not mentioned. When the electrician was finished, he asked the customer to turn on his radio.

The radio was almost silent, S-1 to S-2, which is about as good as it can get—until he turned on the fluorescent light. The electrician said he would replace the old ballast light with a new light at no cost.

The customer turned to me and stated, "I guess I am not as smart as I thought I was." The customer seemed embarrassed by his own behavior and appeared to be very humbled by the entire experience.

He said to me, "I am glad I am not you."

"Why?" I asked.

He answered, "I would not have handled things nearly as professional, and there is a good chance I would still be trying to operate with a noisy radio."

There was no need for me to say anything, as he then realized how hard I had tried to help him. My goal was to help this highly decorated WW2 veteran to be able to spend his retirement doing what he enjoyed.

I thanked him for his time, and we shook hands. I called him from time to time to see how things were going. Each time, he was very nice to talk with, and he enjoyed his radio for many more years. Once in a while, I would need to resolve an issue that popped up for him, but he always had full confidence it would be resolved quickly, and it was.

We stayed in touch until I got a call that he had passed away at the grand old age of 94. Through the years, he passed on a lot of knowledge to me about ham radio, the war, and life in general.

He was a rough old fellow that many may have seen as a stern old guy who was no-nonsense, and some may have even considered him rude and too outspoken. What I saw was an old fellow who had his flaws like anyone else (including myself), who had served his country honorably, and who deserved to live during his retirement in any manner he wished.

I was proud to have helped him resolve his radio problem, and I am still proud to have called him my friend.

This segment is dedicated to those ham radio operators who served their country and continue to serve their country, their state, and their local area by being ready when disaster strikes.

Ham operators have played a vital role in logistics for every major hurricane and have helped save countless lives. Ham operators provide this service at no charge, as that is what they enjoy doing. Ham operators are a vastly overlooked resource for utilities and other local agencies.

Most Cities and towns have a local ham radio club. I would strongly suggest that any Utility get involved with these operators, as they hold a vast amount of knowledge that will be there when needed. I know from experience what I have learned from numerous members through the years.

You must be careful, though, or they might recruit you to cook at their yearly hamfest meeting. In all seriousness, though, I really enjoyed every moment of the event, and I value the friendships that were made. Special thanks to all the ham clubs around the country that are ready to help anytime they are needed. What a great group of men and women.

Residential Open Neutral

I WAS CONTACTED BY ONE of the Utility system operators due to a possible ongoing neutral issue at a residential home. Two different troubleman linemen had been to the location over the past three days, and all checked normal with what is known as *the Beast*.

The Beast is a device that is inserted into the socket of residential meter-can jaws, just as you would insert it into a normal meter. Then you can apply load to one leg at a time while you watch the voltage, and you can toggle back and forth between the two legs.

The load is about 12 amps. If a bad neutral is present, the voltage will read high on one leg and low on the other, as the neutral keeps the load balanced. But without a neutral, the voltage can be up to 240 volts on one leg and 0 on the other, which can cause a lot of problems.

I contacted both linemen who had been to the location. After they had given me a general overview of all factors involved, we discussed what they had done.

Each lineman said they verified proper voltage and changed connections on the Utility side. Both linemen had talked with the customer who had stated that the problem of lights getting very bright and then very dim had only been happening from time to time but recently was happening much more often.

Each lineman had checked for a bad neutral with the Beast, and each stated there was a small amount of voltage fluctuation but not enough to indicate a neutral issue. Both linemen felt the Utility side had no issues and attributed the small voltage fluctuation of the Beast to normal voltage flicker due to wire size and distance from the transformer.

Both linemen had advised the customer to seek an electrician to look for possible issues on her side of the meter. The customer contacted an electrician after the second lineman had advised her to.

The customer then called customer service to state that the problem was still happening, and that her electrician had stated there were no problems on the customer side of the meter. That is when I was contacted to get involved and verify whether a problem did or did not exist on the Utility side.

I called one of the linemen who had recently been to the customer location and asked him to meet me there to do some troubleshooting.

At this point, it had been almost a week, and the customer had called in twice and explained symptoms of a bad neutral connection. Both times, our linemen had found no issues and had changed connections at the weatherhead of the trailer pole as well as at the transformer.

The customer had again called an electrician to verify her side of the meter, and the electrician stated there were still no issues on the customer side.

At this point, this was becoming a possible liability issue for the Utility. If anything had been overlooked on the Utility side, the customer could have a valid damage claim. I met with the customer and explained that a recorder would be installed at the meter to monitor voltage and current for one day.

The next morning, I went to the customer's home and recovered the recorder. When I downloaded the data back at the office and looked at everything involved, I was very surprised to see the data showed an intermittent bad neutral on the Utility side.

I called the lineman and asked him to meet me at the customer's location again because the recorder had shown an issue on the Utility side. The lineman's initial response was, "That can't be correct." He said he had checked and verified everything.

I also called the customer and informed her of what I had found and asked that she contact her electrician to meet with me on site so we could check all possible aspects and everyone would be present.

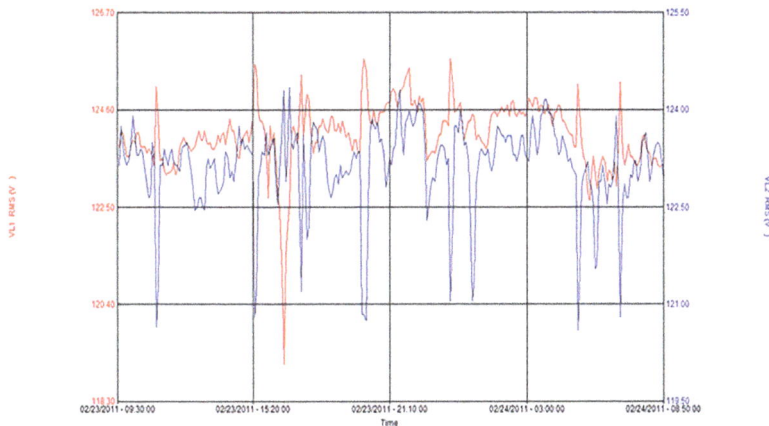

The above graph shows the intermittent issue. The opposing voltages are a good indicator of a bad neutral.

The customer's electrician and I met the lineman on site. I explained that my recorder had indicated an issue on the Utility side, and we were going to replace and upgrade the 125' #2 overhead triplex service cable with 125' of 1/0 overhead triplex.

This was the only element that had not been changed; the linemen had changed connections at the weatherhead and transformer twice.

The transformer was not an issue, as the other customer that fed from the same transformer was having no problems. The service was quickly replaced, and everyone felt the issue had been resolved, most likely due to a bad spot in the service neutral.

I checked the service after everyone had left, and I found no sign of any problem with any aspect of the service. I told the customer I was going to reinstall the recorder to verify the issue had been resolved. I arrived the next morning to recover the recorder and was told by the customer that the issue was still happening.

I took the recorder back to the office and looked at the data, and it still showed the problem to be on the Utility side. I was not sure where to go from there, as every aspect had been replaced on the Utility side.

I called everyone to get them to meet me again at the location. The electrician and the lineman were not happy that they were still involved in this, and each of them questioned whether my recorder was providing proper information.

I maintained my equipment very well, and I trusted it, so I assured everyone the equipment was correct. While they were talking, I stopped and just looked at all involved aspects and realized that was what I should have done earlier.

I realized that I may have overlooked a possible cause. I walked to where our lineman and the customer's electrician were standing and asked the electrician, "Are you sure there are no issues on your side?"

He replied that he had checked everything, and no issues were found.

I knew everything had been replaced on the Utility side, so I told our lineman to disconnect the service from the transformer and then disconnect it from the weatherhead. I asked the electrician if he would dismantle the standpipe from the meter can so we could see the wire from the top side of the meter to the connection point of the weatherhead.

Both of them looked at me at the same time and agreed that had to be where the issue was. I had overlooked the standpipe as a possible option of where the problem could be. Although the standpipe is on the customer's side, and is the customer's responsibility, the stand-pipe wire will show as a problem on the Utility side due to it being prior to the top side of the meter.

The standpipe was removed and—*wow*! The neutral was covered with corrosion, and it broke in two pieces as we removed it from the pipe. We were all amazed and relieved that the issue was finally found and would be resolved.

This problem had to have happened when the service was installed many years earlier. Most likely when the wire was put in the pipe, it was scraped, and over the years oxidized, and over many years reached a point of failure.

The electrician replaced the wire, and the issue was resolved. And yes, I did install a recorder one more time to be able to provide the customer with before-and-after voltage and current profiles of her home.

All of us agreed that we had learned from this issue. I know I will never forget this one. It was one of the instances that taught me to look at the big picture and that sometimes you must slow down and look much deeper than you normally would need to.

I recovered the recorder the next morning and took it back to the office. The results showed good voltage with no issues. I met with the customer and explained the profile to her.

The customer was very grateful we had been able to resolve the problem. Sometimes it is very difficult to trust what your equipment is telling you, but you have to learn to trust your equipment, at least until it shows you verifiably incorrect data.

Interpreting Recorder Data

INTERPRETING THE DATA COLLECTED FROM your various power quality devices is the key to resolving or identifying potential issues. This aspect of power quality is often not looked at as an important process.

I say that because I have known many utilities to just look at voltage they are providing, and if voltage is within accepted guidelines, they move on.

I am not going to get too involved in the recorder data in this introductory book, but I do want to mention there is much more to properly interpreting data than just looking at the voltage.

I highly suggest anyone responsible for interpreting recorder data consider taking power quality classes, preferably from Cardinal Power LLC, as I know my classes are quality classes that will build your knowledge base in power quality and reliability.

But no matter where you decide to seek your training from, make sure the instructor is well versed in power quality and reliability, and not just giving you a sales pitch for their product.

I would highly suggest you buy the Dranetz *Handbook of Power Signatures*. The cost is usually less than $100. This book will be a main reference when learning to read sinewave signatures, which will come later, but I would suggest that you get very familiar with this book early in your learning.

The *Power Signatures* book has many sinewave signature pictures that represent different types of Utility events. Each Utility event is like a fingerprint; they each have their own unique signature.

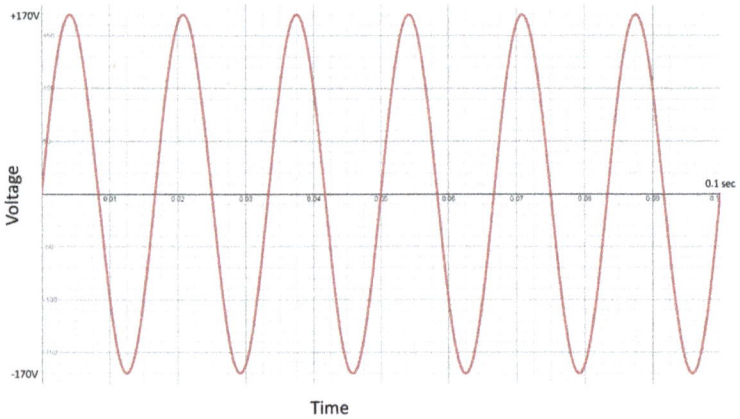

I will provide much more detail in my next book concerning sinewave signatures, but for now it is important to attend power quality and reliability classes, preferably by Cardinal Power LLC, if you have not already done so.

Root Causes Are Often Overlooked

THERE IS A WIDE RANGE of different problems that can cause equipment to malfunction. Some of those causes are poor wiring, loose connections, overloading, other loads from the same feeder panel. . .the list can go on and on.

Many of the newer technologies produce new challenges for the user, such as injected noise onto data cables which can cause data errors. Many customers do not consider needing shielded cables for the newer technologies. If the data cables are in close proximity to power or other cables, that may produce EMF (electromagnetic field) and cause data errors.

If fiber optic lines are available for data transfer, I suggest using fiber, as many of the issues do not exist with fiber.

Some of the most overlooked problems are due to the environments of the devices. Many, many times, I have seen a case in which an equipment problem is tracked to a bad connection. The connection is then repaired or replaced, and nothing else is considered.

Yes, there are times when connections just go bad due to age or lack of maintenance, etc. However, in many cases there is a reason for the loose connection beyond the obvious. This is why in a previous chapter I mentioned you should take your time and look at everything involved.

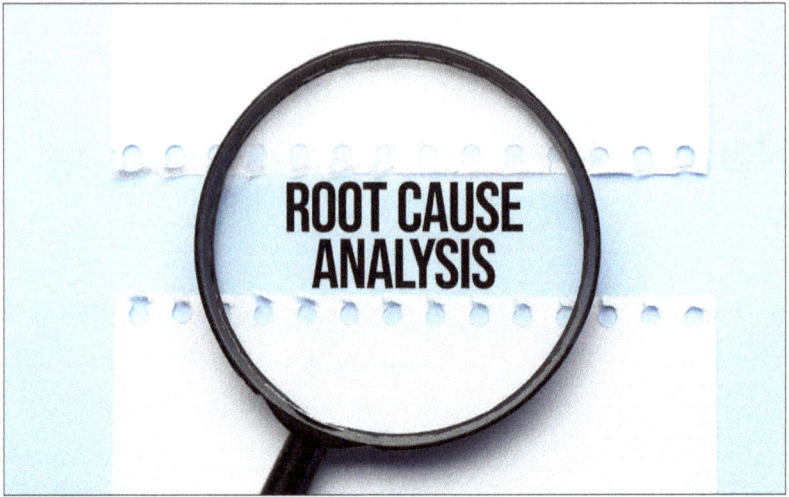

First, consider if the area where this connection was found is a common area for other connection-related problems, more so than in other areas of your business.

Next, consider the environment. For example, if your business or home is near a railroad track where train traffic is frequent, the vibration of the trains will cause panel connections to loosen over time.

I have found this many times throughout my career, and each time the customer or their electrician admitted, "I never considered that as a possible cause."

Many customers have rail docks for their business to send or receive products. These docks have lighting, in most cases, as well as many other electric needs, such as offices, cranes, etc.

If you have equipment in this environment, it is a good idea to provide general maintenance more often than you would in an area that does not have the vibrations of the train.

There are several other areas that can also be affected by very similar environments, such as docks where semi-trucks bump the dock when backing in to be loaded or unloaded.

Over time, the bumps can cause connections in the general area to loosen; therefore, I would also have an enhanced maintenance schedule.

Another location to consider is warehousing. If forklifts are being driven in the building, consider the vibrations in that area caused by those.

Bowling alleys are notorious for having bad connections due to the constant crashing of the bowling pins and the vibrations they create.

In general, if you can feel the vibrations in your building, so can your panels. If you are investigating a voltage issue, just keep in mind some of the items I have mentioned. Considering these factors could save you a lot of time in your search.

Over the years, I have discussed train vibrations and their effects with many people, but one of the most interesting of those was Mike, a train conductor's son who attended one of my **Power Quality & Reliability** training classes.

When I started discussing the vibration issues of trains and the problems they can create with electric panels, he was very eager to tell the class a story from his dad, who had been a train conductor for 31 years.

Mike told the class his dad had mentioned to him many times that cars delivered by rail have a lot of loose bolts and screws due to the constant vibration during their journey.

Mike mentioned that a major air conditioning company had also been having problems with their shipments of air conditioner units because the bolts and screws were falling out of them during the transport.

Initially, the company thought the units were just not being assembled properly and continued to ship by rail. After verifying the units were properly assembled, however, the company found that the problems continued.

Due to many customer complaints and product issues, the company decided to not ship any units by rail for three months and just use over-the-road trucks instead to deliver their products.

Mike stated that the problems of bolts, nuts, and screws falling off, and of components being loose, were resolved.

To this day, the company's products are not shipped by rail due to the effects of the constant vibration.

Take All Outages Seriously

I HAVE HEARD SOME SAY that a residential outage is not as important as a commercial or industrial facility outage.

This may be true from a monetary standpoint, but let us take a moment to consider all possible aspects involved. I agree that in *most* cases, an outage to a residential home is an inconvenience.

The homeowner may not get his or her usual cup of coffee, or they may need to stop by and get breakfast on their way to work. To consider it a huge monetary loss is usually not the case unless the outage is for an extended period of time.

On the other hand, an outage at an industrial or commercial facility can have much greater financial impact. Outages to these facilities can cost a lot of money in lost product as well as unproductive man hours when you have many employees standing around, doing nothing.

Safety is another concern in commercial and industrial facility outages. When an outage occurs, many of the buildings are enclosed and all goes dark. Imagine what could happen if a forklift driver has a pallet of merchandise 25 feet in the air when the lights go out. That's right—a huge safety issue.

Depending on the type of lighting, the lights may take several minutes to come back on once power is restored. Most companies are now using LED lighting that does not have to restrike; therefore, when power is restored, the lights are also restored.

Let's get back to the idea that a residential outage is not nearly as important as a large business outage.

Stating that one type of outage is more important than the other is a ridiculous statement, in my opinion. Circumstances will dictate the importance of an outage.

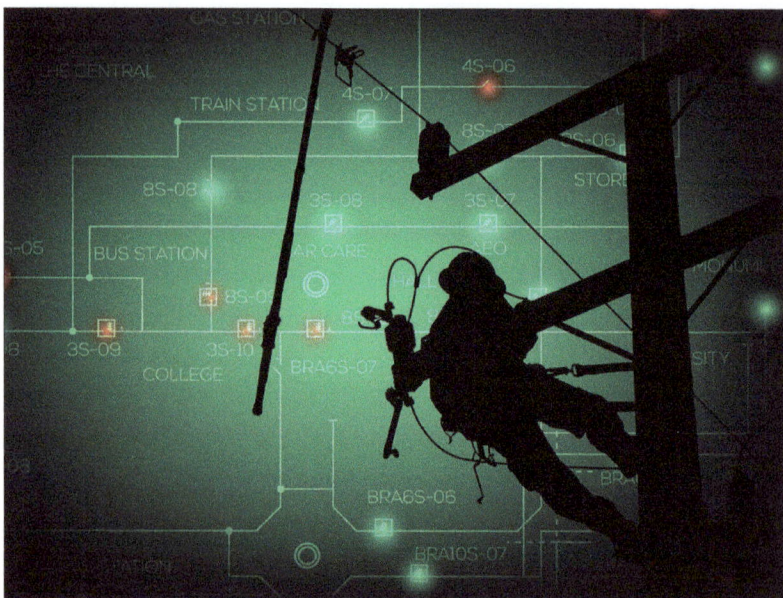

Through the years, I have had many residential customers contact me and express their concerns about outages. Some of these customers had medical devices that needed power. I explained to them an online UPS may be a good idea because no Utility can provide perfect power all the time, and no Utility can promise not to have outages. Many things, of course, are beyond the Utility's control.

To provide the best and most reliable power quality possible, I would always look at all factors involved, but there is one case etched in my memory that left me no doubt a residential outage can be of the highest priority.

I was contacted by phone early one morning just after I had arrived in my office. The customer stated she had contacted customer service and they had given her my direct number.

Usually, customer service did not give out my number but instead would provide me the customer's contact information so I could return their call, so I knew this situation must be very different.

When the lady asked if I had a little while to discuss her concerns, I assured her she could take as much time as she felt she needed. The lady then explained that her twelve-year-old daughter had terminal, stage-four cancer and had recently been sent home with a lot of electrical equipment needed for the child's pain management, medication, bed elevation, etc. Without power, therefore, she was in a lot of pain. The mother stated there had recently been a few outages that affected her neighborhood and anything that I could do to help keep the power on to her home would be greatly appreciated.

I explained that I would be out to her neighborhood as soon as I had gotten all the outage data together, and that I would look at all factors involved in her area to try and avoid any outages.

I did explain the benefits of an online UPS, but she told me her family had checked into that already and found they would need several devices or one large one that would cost a few thousand dollars, which they did not have.

I collected the outage data and learned the recent outages were caused by an osprey trying to build a nest on a nearby switch. I went to the customer's location and explained I would be adding bird guards and anything that may help prevent outages, and that I had a crew on the way to implement the upgrades.

The customer thanked me and was very happy with the response she had gotten from the Utility. I met with the crew and installed a fake owl, as a deterrent, and some e-tape on the bare connections (the jumpers already were insulated).

All was good for a few days, and then I got a call that they'd had another outage. I went to the switch and found an osprey sitting on top of the owl. There were several small tree limbs at the top of the switch that could fall and cause another outage and maybe even burn up the switch. I knew I had to come up with something to keep this bird from causing outages. I went back to the transmission and distribution yard reclaim area, where a lot of used parts were available, and looked to find what might work.

As anyone who works at a Utility and has had to try and keep ospreys from building a nest on their equipment during nesting season knows, ospreys are very determined to build where they want.

You can even provide a nesting platform for them, and some will use it and some will not. This one would not use the nesting platform that was less than 100 yards away.

I saw a friend of mine at the yard, and we both discussed the issue and started searching for anything we thought might work. We came up with a plan to use rebar, PVC pipe, and Guthrie Guards.

I had a crew meet with me at the switch to get their thoughts on whether our idea might be better than what was there, as the owl clearly was not working.

We bent the rebar and heated the PVC pipe over it to help insulate it. The rebar was inserted into each end of the crossarm and was completely covered with PVC and e-tape.

The PVC portion that was above the energized conductors was left where it could spin. I also had Guthrie Guards installed around the PVC pipe. This design would prevent the bird from landing because the PVC would spin and the Guthrie Guards would give a small electric shock due to the induction from the nearby energized conductors.

The bird guard worked perfectly, and there were no further outages during the young girl's remaining months.

This bird guard is still in place and is still working well. Many people at the Utility state it is the best bird guard design they have.

I have often thought of this young lady when I was working to improve reliability. Whether a residential outage is as important—or more important—as a large commercial or industrial outage is always dependent upon the circumstances involved.

Utilities & New Technologies

NEW TECHNOLOGIES ARE ALL AROUND us, from drones being used to inspect circuits, to self-healing circuits. Keeping up with the new technology that is available can be a job, not to mention the learning curve needed to have a better understanding of them.

A general understanding of the newer technologies is needed to be able to determine whether the new technologies would be beneficial for your Utility.

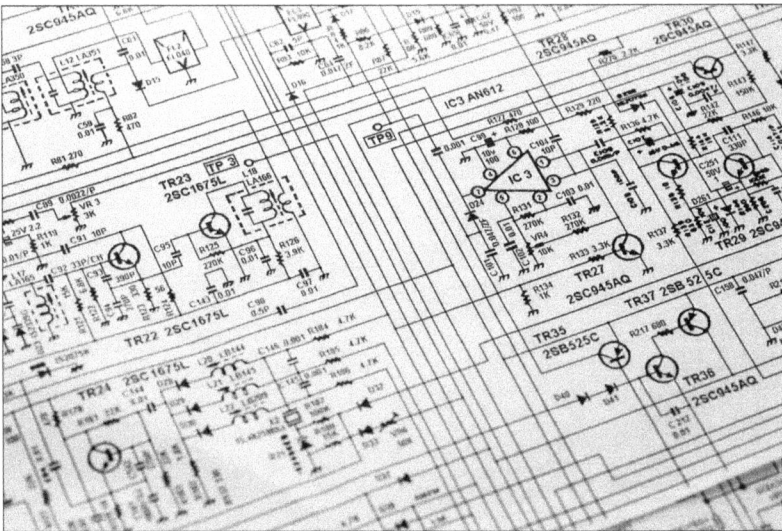

Change is very difficult for some people, including me. I have a mindset of "if it's not broke, don't fix it," but through the years I have slowly adapted to what I consider a better way of looking at change.

Change can be very interesting and can even increase teamwork and boost morale. Implementing pilot projects to make sure the new technology is a good fit for your Utility is usually a good idea.

I recently added a new class to my training called **Utilities & New Technologies**. I created this class at the request of several people who asked pertinent questions about new technology and the impact it may have on utilities.

Cardinal Power Training Programs

IF YOU HAVE FOUND THIS book interesting and would like to have Cardinal Power LLC provide training at your facility, please contact me at mark@cardinalpq.com or call 931-310-3077.

A complete list of classes can be found on the Cardinal Power Facebook page, or you can email me to request a complete Training Package with all the details.

Added Benefits of Our Classes:

- I come to your location. This saves you money on company travel costs, meals, and many other possible expenses, which in most cases would be more expensive than the total cost of my training.

- I also provide online/virtual training.

- I provide PDH/CEU credits and certificates for all who attend my classes.

- I try to limit classes to 75 students or less, which provides better class participation.

CARDINAL POWER

In Closing

CHANGE IS INEVITABLE, AS TECHNOLOGY is going to keep improving and the learning curve will only get steeper. Making sure your employees have at least a general understanding of the causes and effects of the newer technologies could have a huge impact on how well they are able to serve the Utility as well as their customer base.

I look forward to seeing you soon.

Remember: *Knowledge is "Real Power"*

Best Regards,
Mark A. Shirah, CPQ
V.P. of Operations
Cardinal Power LLC
mark@cardinalpq.com

About the Author

I HAVE BEEN INVOLVED IN the field of power quality and reliability for over 30 years, with five years' lineman experience and over 25 years of pure power quality and reliability work. I am one of only 181 Certified Power Quality (CPQ) experts in the world as of this writing.

I have resolved over 7,000 power quality or reliability issues for large industrial complexes or plants like Publix Industrial Complex, Pepperidge Farms, Rooms to Go, Lockheed Martin, University of Florida, Mission Foods, Amazon, Federal Mogul, various international airports, many small businesses, and residential customers.

I am V.P. of Operations for Cardinal Power LLC, which is a one-stop shop for any power quality monitoring, training, advice, engineering services, and equipment needs.

I have been married for 40 years to my wife, Debbie, who has been an enormous support in all aspects of our business and personal life, including the writing and production of this book. We have one son (Cody) who is also highly involved in power quality and reliability. He, along with many friends and family members, has been a huge inspiration in the writing of this book.

My family and I are Christians who do our best to pass the Word of God to others. I have no doubt God has kept me safe through the many years, and I realize that without God I am nothing. But with Christ, I can do all things.